THE OCEANS

SALLY RIDE
SCIENCE

Our Changing Climate
THE OCEANS

Contents

OCEANS RULE!

You've heard the news—the world is getting warmer. Even the very largest of ecosystems, our oceans, are feeling the heat. In fact, when it comes to understanding how Earth is responding to climate change and what it means for our future, scientists look to the oceans for important clues.

Without a doubt, oceans are major players in keeping our planet healthy. They cover more than 70 percent of the globe. They are home to more than 80 percent of life on Earth—from seaweed to sharks. They store heat. They regulate our weather. They provide much of our food. Many of our cities hug their shores. And with a system this huge, once change is under way, you can't just put the brakes on. For generations to come, we're going to feel the effects of what's happening in our oceans today.

The Facts

Volume	1,332 million cubic kilometers (320 million cubic miles)
Average depth	3,682 meters (2.3 miles)
Temperature	minus 2°C to 29°C (28°F to 84°F)
Average Salinity	3.5 % (mostly sodium chloride—NaCl)
Explored	less than 5%

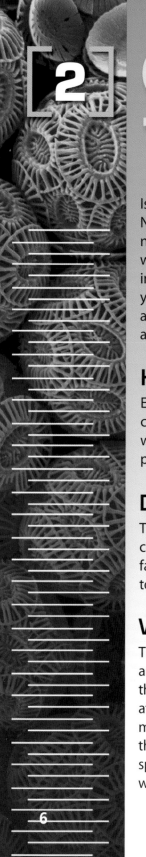

CHANGE IN THE AIR

Is climate just a fancy word for weather? No. Climate is related to weather, but it's not the same. Weather is what you see when you look out the window. Climate in your hometown is the average weather you can expect where you live. But you can also talk about the climate of a country or a continent or the whole planet.

Heating Up

Earth is getting warmer. That means climates around the world are changing. They're not all changing at the same rate or in the same way—but they're all changing. And that's affecting everything on our planet in one way or another.

Daily Dose of Sunshine

That big yellow ball in the sky, the Sun, powers our climate. The Sun constantly emits energy in all directions. Fortunately, a small part of it falls on Earth. Sunlight provides the light and heat that we depend on to live.

Warming Our World

The sunlight that strikes our oceans and land is absorbed at the surface and warms the planet. The warm surface then tries to cool off by radiating the heat back toward space. If this heat could make it out through the atmosphere as easily as the sunlight makes it in, our planet would be much colder than it is. But not so fast! A few gases in the atmosphere— the greenhouse gases—absorb some of the heat before it escapes into space. They trap the heat and make our planet warmer than it otherwise would be. Yes, this is the greenhouse effect (right).

Only 1 Percent

Not all gases are greenhouse gases. In fact, about 99 percent of our air is made of gases that are *not*—oxygen and nitrogen! But without the other 1 percent, there would be no greenhouse effect on our planet. You might think that would be a good thing. Think again.

Earth's atmosphere is mostly nitrogen and oxygen.

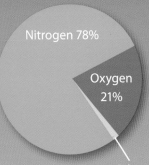

Nitrogen 78%

Oxygen 21%

Other gases 1%
(including water, carbon dioxide, ozone, and methane)

Hello, Greenhouse Gases

The most important greenhouse gases are water vapor, carbon dioxide, and methane. They're nothing new. They were floating in Earth's air long before there were people on the planet. And though they're only a tiny percentage of our air, those few molecules provide a greenhouse effect that warms Earth. If there were no water vapor or carbon dioxide in our air, Earth would be about 33°C (59°F) colder! Our planet would be one big ice ball.

Greenhouse Gone Gonzo

If the greenhouse gases in our air keep Earth from freezing, what's wrong with adding more of them? Those gases that we're sending into the air are causing even more warming. And that's affecting our whole planet.

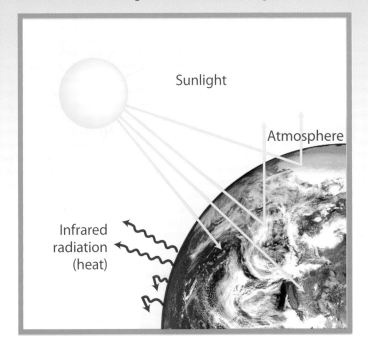

Sunlight

Atmosphere

Infrared radiation (heat)

As carbon dioxide and other greenhouse gases from car and factory fumes build up in the air, our planet is getting warmer and warmer. Uh-oh. Greenhouse overload.

Back to the Future

Climate change is nothing new. Over Earth's long history there have been cooler times, like the ice age that happened 15,000 years ago. And there have been warmer times, like the tropical dinosaur days that ended 65 million years ago. In the past, these climate changes were usually triggered by natural shifts in the Earth-Sun orbit, or by changes in the amount of sunlight reaching Earth. That's not the case today.

Who, Us?

This time, climate change is different. Humans are the cause. How did we do that? We've changed the atmosphere . . . much faster than it's ever been changed before. Many of the things we do—driving cars, flying in planes, lighting our cities, making things in factories—add greenhouse gases to the air. And we're adding lots of them.

How Do They Know?

Aloha, CO_2

Before 1958, no one knew how much carbon dioxide was in the atmosphere. That year, a young scientist named Charles Keeling set up a monitoring station near the top of Mauna Loa, the largest volcano in Hawaii, to find out. He measured the amount of carbon dioxide in the air continuously for many years. His measurements were used to create one of the most famous graphs in science.

The Keeling Curve (right) shows that the amount of carbon dioxide in the air has gone up every year. When the measurements started, there were about 315 molecules of carbon dioxide out of every 1 million molecules of air—or parts per million (ppm). Today, there are around 400 ppm of carbon dioxide! This is a huge increase in a short period.

315 ppm
1958

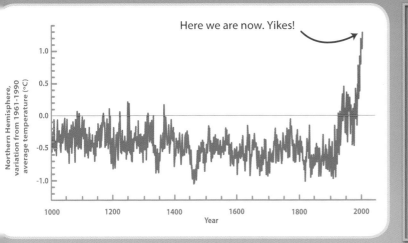

Here we are now. Yikes!

Northern Hemisphere, variation from 1961-1990 average temperature (°C)

Year

Compared to daily shifts in weather, climate change is subtle and hard to measure. It took scientists years to be sure it was real. But it is. In the last century, Earth's climate has warmed about 0.8°C (1.5°F). That may not sound like much, but it's the fastest our planet's global average temperature has changed in 1,000 years.

397 ppm
2013

341 ppm
1982

What causes those zigzags? Plants! The graph goes down in the spring, when plants in the Northern Hemisphere grow and suck in carbon dioxide as part of photosynthesis. It goes back up in the fall, when leaves drop from the trees and many plants go dormant. The squiggles show Earth "breathing." But Earth's breathing is the opposite of ours—it inhales carbon dioxide and exhales oxygen.

Warming Signs

So it's getting warmer. What's the big deal? Well, scientists have already measured many changes all around the planet. The oceans are warmer. Glaciers on mountains and ice caps are shriveling. There's more rain in the northeastern U.S., and storms are more intense. There's less rain in the parched southwestern U.S. And ecosystems everywhere are changing.

9

[3] WATER, WATER EVERYWHERE

Take a good look at Earth. Whoever gave it *that* name? It's blue—definitely the water planet. More than 70 percent of Earth's surface is covered by sparkling blue salt water. The oceans affect the whole world—from the air to the land to living things, including us.

It's Raining, It's Pouring

The story of our seas started more than 4 billion years ago. Shortly after Earth formed, water couldn't exist on its scorching hot surface. But eventually, the young Earth cooled enough for water vapor in the air to form clouds. That's when the rains came. Rain poured down for centuries. It collected in low-lying lands, gradually filling up into mighty oceans.

Cast of Characters

Although it really is one vast ocean, we usually break it down this way, from largest to smallest.

Pacific—Covers more area than all seven continents combined—about one-third of Earth's surface

Atlantic—Includes the Mediterranean, Baltic, and Caribbean seas

PACIFIC OCEAN

A1

Hermit crab

It's Alive

The oceans are home to 80 percent of all life on our planet. Together they make up the world's largest ecosystem. In fact, 99 percent of the actual living space on Earth is under water. Earth's earliest recognizable life forms—tiny single cells—appeared in the oceans about 3.5 billion years ago. For millennia, microbes like these were the only life on Earth. Then, about 800 million years ago, more complex plants and animals evolved. Green algae floated in sunlit waters, and small, soft-bodied animals left their tracks on the seafloor.

Being There
Too Many to Count

Marine biologists estimate that there are millions and millions of species of plants, animals, and microbes living in our oceans. Just how many is anyone's guess, since 95 percent of the ocean world is still unexplored! Marine life is spectacularly diverse—from the tiniest microscopic zooplankton to whales, the largest animals ever to live on Earth.

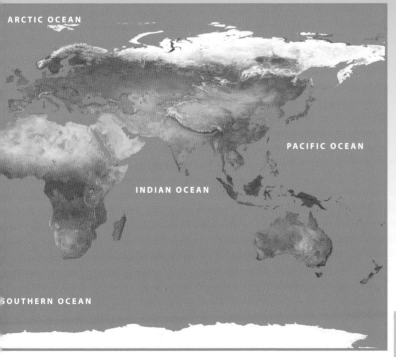

ARCTIC OCEAN

PACIFIC OCEAN

INDIAN OCEAN

SOUTHERN OCEAN

Indian—Includes the Persian Gulf and Red Sea; home to a fish smaller than a fingernail

Southern—Home to some of the richest feeding grounds for fish and marine mammals

Arctic—Covered by ice most of the year; almost completely surrounded by land, including parts of Greenland, Alaska, Canada, Russia, and Scandinavia

Giant clam

TEMPERATURES RISING

Just like a backyard kiddie pool slowly heating up on a sunny afternoon, our oceans have gradually been getting warmer as air temperatures rise around the world.

One Degree? What's the Big Deal?

Over the last century, Earth's air temperature has risen almost a full degree Celsius. That may not sound like much, but it means that the world is warmer now than at any time in at least 1,000 years! So what? Well . . . our planet actually operates within a pretty small temperature range. During the last ice age, average temperatures were only about 7°C (13°F) cooler than today. So small changes can make a big difference! Over time, a little extra heat adds up. A small change in temperature can cause enormous environmental changes.

Less Than 1 Degree?

Using research ships and buoys, oceanographers have collected millions of water temperature readings. They've learned that worldwide, the average temperature of the oceans has risen 0.04°C (0.07°F) in just the last fifty years. Just 0.04°C!? Can such a little bit matter? Yes!

Our oceans absorb 70 percent of the sunlight striking them. But because it takes so much energy to heat water, oceans warm more slowly than air.

Harming Habitats

And remember that 0.04°C (0.07°F) is averaged globally. Some parts of the ocean are warmer than others. In tropical waters, average temperatures have risen as much as 1.7°C (3.7°F), and many coral reefs are struggling to survive. In the Arctic, waters are warming even more and sea ice is melting faster than ever before. This has altered food sources all sorts of marine animals depend on.

As Arctic sea ice melts, life gets tough for animals that depend on it, such as this Pacific walrus.

Can You Say "Thank You"?

Listen—as bad as scientists say global warming is already, it would be much worse if it weren't for the oceans. They've been busy sponging up heat—bailing us out big-time. Over the last 50 years, the oceans have absorbed more than 80 percent of Earth's extra heat trapped by greenhouse gases. That's 14 times more than was absorbed by the atmosphere. If all that stored heat were suddenly released into the air, air temperatures would rise by about 22°C (40°F)! Do the math—add that to today's temperature!

The World's Biggest Science Experiment

The jury is in. Human activity is affecting our planet like nothing Earth has ever experienced before. It's impossible to foresee exactly what the future may hold for us. But we know that the changes we're seeing today are only the beginning. Scientists can make important predictions because, in many ways, water warmed in the ocean reacts the same as water warmed in any lab.

H₂O Chem

Even out there in the deep blue sea, the basic rules of chemistry hold true. What happens when you heat water? Apply these concepts on a worldwide scale and you get megaconsequences.

RULE #1

Water expands when it's heated.

As water warms up, water molecules move around more—they twist and turn more quickly—so they move farther apart. This means the same amount of water takes up more space. Centuries ago, the first thermometers were simply water-filled glass tubes that recorded temperature changes from the changing water levels. When oceans warm, they take up more space, too. The result? Sea levels around the world are rising.

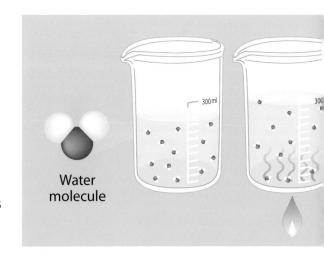

Water molecule

RULE #2

Warm water weighs less than cold water.

Warm water molecules move around more and spread farther apart. This means there are fewer water molecules in a cup of warm water than in a cup of cold water. The result? Warmer water is less dense and lighter than cooler water. So warm water floats on top of colder water. Out at sea, this can upset the way nutrients mix through the water, hurting ocean plant life.

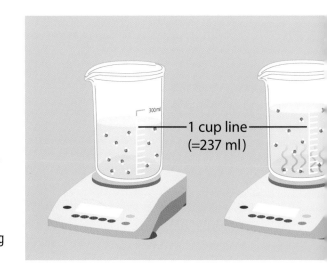

1 cup line (=237 ml)

RULE #3

Heat water enough and it turns to gas, or evaporates.

The water molecules dance around so quickly that they break free and escape into the air. Think steaming teakettle. This happens in the oceans, too. And water evaporates more quickly as temperatures rise. More water vapor escapes into the air. This is affecting rainfall patterns around the world.

RULE #4

Heat ice and it melts.

This is what's happening right now in the Arctic—where the effects of global warming are most visible. Over the last century, Arctic temperatures have increased at twice the global rate, making it harder for sea ice to form. Having less sea ice is affecting ecosystems all across the Arctic.

Vanishing Act

Measurements taken by ships, airplanes, and satellites show the cap of sea ice floating on the Arctic Ocean is growing smaller and thinner. Scientists are shocked at just how quickly the sea ice is melting. Since record keeping began in 1978, summer sea ice has been shrinking by about 60 million square kilometers (23 million square miles) each year. That's like losing half of Ohio every summer. It's fading faster than scientists ever thought possible. Experts predict the Arctic summertime ocean will be completely ice-free by the middle of this century.

A young leopard seal naps on the sea ice.

Polar Bears—Cold, Hard Facts

When sea ice goes missing, animals that have depended on it for centuries are hurt. Up north, polar bears that used to roam over sea ice to hunt are going hungry. Canada's Hudson Bay is now ice-free 3 weeks longer than it was 30 years ago. That's bad news for bears—the polar bear population there has dropped 22 percent.

Penguins—Cold, Hard Facts

In the Antarctic, the number of emperor penguins has shrunk by 50 percent during the past 25 years. Why? With less sea ice, there is less ice algae, a major source of food for tiny, shrimp-like animals called krill. Guess who eats the krill?

Emperor penguins strut across the ice in Antarctica.

Experts Tell Us

Karen Bice

Paleoclimatologist
Woods Hole Oceanographic Institution

Karen is a time traveler. And, like some sci-fi character, she returns from the past with lessons for our future.

To find out what oceans looked like nearly 100 million years ago, Karen studies the fossilized shells of tiny marine microbes called *Foraminifera*. She learns about the environmental conditions that affected how these animals grew, such as temperatures and carbon dioxide levels, by analyzing their shells. This information helps her understand how our oceans may change as Earth warms today. Recently, she discovered that ancient seas were much hotter than anyone thought—as hot as 42°C (108°F). That's hot-tub hot! Knowing this, she believes scientists are underestimating just how warm our oceans could get as carbon dioxide levels continue to rise.

Karen says it's important to look at our planet's history for answers. "Earth is warming right now, and we don't know how warm it's going to get before it stops getting warmer," she says.

Coral Reef Grief

Coral reefs are formed by millions of tiny animals called corals. Each little coral builds a hard, rocky skeleton around itself. As generation after generation of corals builds new layers of rocky skeleton on old layers, a reef is made. The cracks and crevices in the reef provide homes to an amazing array of other creatures. The reefs are teeming with life. Even though coral reefs cover less than 1 percent of the ocean floor, they provide a home for 25 percent of all fish species.

Coral scientist Orla Doherty tracks the health of coral around islands in the South Pacific Ocean.

Whiteout

Many corals are struggling to beat the heat. Most can live only in waters between 18°C and 30°C (64°F and 86°F). When ocean temperatures get too high, corals lose the colorful algae that live inside them. This causes them to turn white, or bleach. Without the nutrients provided by the algae they host, corals can starve. According to scientists, about 25 percent of the world's coral reefs have already been destroyed.

Great Barrier Bye-Bye

The Great Barrier Reef off the northeast coast of Australia is like a vast undersea nation. Its citizens come in a dazzling array of colors, and in all shapes and sizes. The many species found there include some of the most beautiful in the sea—angelfish, butterfly fish, and dolphins; minke whales, manta rays, sea turtles, and sea urchins. All depend on precious corals for food and shelter. Warming waters are wreaking havoc on this astonishing habitat.

In 2002 more than 60 percent of the corals suffered bleaching, turning part of the reef into a boneyard of skeletons.

Turtles—Feeling the Heat

In a rapidly changing world, animals with rock-steady habits suffer most. For generations, sea turtles such as hawksbills (left) and loggerheads have swum thousands of kilometers across the ocean to lay their eggs on the same stretch of sand where they were born. But as sea levels rise, shorelines are shrinking. This leaves less beach for turtles to lay their eggs. Warmer waters are also damaging coral reefs where turtles find food. Scientists wonder—will turtles be able to adapt fast enough to survive?

Warm sand may feel good between your toes, but it is changing the gender of turtle hatchlings. Recently in Florida, 90 percent of all loggerhead babies born were female.

Measuring a Hidden World

It's easier to find out what's happening on the Moon than it is to get an accurate picture of what's really going on in the world's oceans. Here are just some of the tools scientists use to crack undersea mysteries.

Spaced Out

In a single hour, orbiting satellites can scan an entire ocean. Scientists use these pictures to measure sea level and near-surface conditions such as temperature, plankton abundance, and precipitation.

That's a Good Buoy!

Some buoys are anchored, collecting environmental conditions from a particular location. Others are cast adrift, like those in the Argo Project, a program deploying 4,000 floats that travel through the oceans collecting information about water temperature, salinity, and currents. This information is then transmitted wirelessly by satellite to government agencies and universities.

Take My Picture, Please

A Video Plankton Recorder, or VPR, can be towed across the ocean, collecting digital pictures of microscopic organisms to help measure populations and identify species.

Cool Emerging Tech!

Automated undersea gliders are fleets of torpedo-shaped robots gliding like schools of fish across the ocean, gathering data from the surface to the seafloor.

Nothing but Net!

It's definitely not all high-tech. Oceanographers spend plenty of time sloshing around in the wet and cold, pulling up nets to get accurate counts of plankton and fish species. Even the most high-tech pictures from space need real-life confirmation.

Sediment Cores

The fossils of ancient sea animals tell a lot about our prehistoric oceans. Scientists drill them up from the seafloor, and sometimes from spots high and dry like the American West, where an ocean used to be. (It's hard to imagine, but 150 million years ago, Wyoming was under water!) The chemical makeup of this ancient marine dust reveals clues about changing environmental conditions over millions of years.

New Buoy on the Block

Anchored in the Gulf of Alaska, this buoy measures how much carbon dioxide the North Pacific Ocean is soaking up from the air. It also measures ocean acidity—a growing threat to marine ecosystems.

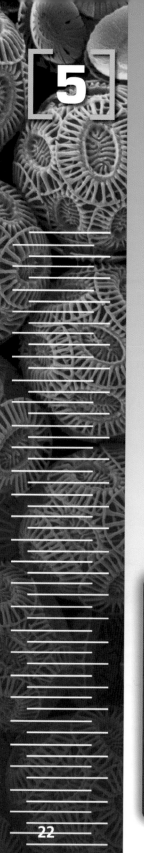

[5] SEAS ON THE RISE

As temperatures creep up around the world, step aside! Seas are on the rise. Warmer oceans need more elbow room. As their volume increases with expanding warmer water and melting glaciers, oceans are rising.

Oceanfront Homes

Right now, about one out of every seven people on our planet lives along a flat coastline. New York City, Tokyo, Rio de Janeiro, and Mumbai—these are just some of the major cities that are located along coasts. More than a billion people live with the ocean at their doorsteps. As oceans rise, more and more people are finding themselves in harm's way.

Rio de Janeiro, Brazil, sits at sea level on the shore of the Atlantic.

How Do They Know?

How Much so Far?

Tide gauges on piers measure the rise and fall of the oceans. They show that sea levels rose at an average rate of nearly 2 millimeters (0.08 inches) a year between 1961 and 2003. Satellites in space are even more accurate and can cover the whole globe. They show that between 1993 and 2003, the rate sped up to 3 millimeters (0.1 inches) per year. That may not seem like much, but . . . there's no end in sight. Experts say that during this century, the ocean will swell anywhere from 18 to 59 centimeters (7 to 23 inches). Better grab your wet suit.

Climate Refugees

The low-lying island nations of Kiribati and Tuvalu—only 5 meters (16 feet) above sea level—have been losing ground fast. And if sea levels rise 1 meter more, the island nation of the Maldives (right) would be under water. Experts warn that more than 150 million people around the world could end up as climate refugees.

The island city Malé, capital of the Maldives, is barely above sea level.

Lair Today, Gone Tomorrow

It isn't just people losing their homes when waters rise. What about plants and animals? The Sundarbans is the world's largest mangrove forest, spreading across parts of India and Bangladesh. The mudflats and marshes of this wetland are a sanctuary for many rare animals, including Bengal tigers. But rising waters are drowning this precious habitat.

A Bengal tiger wades in the water.

Swimming Houses—for Some

Governments around the world are busy planning for a soggier future. To hold back advancing seas, some cities, like Venice, may build massive moveable dam systems. The Netherlands is promoting "swimming houses"—they float in place when it floods (left).

[6] OCEAN LOCOMOTION

The oceans are always in motion. The salty water rises and falls with the daily tides. Winds kick up waves and drive currents near the ocean surface.

Altered States

The ocean is just one stop in water's endless, shape-shifting life—sometimes ice, sometimes liquid, sometimes gas. This constant shuttling between land and sky is called the water cycle.

Together 4 Ever

The water cycle never stops. Over billions of years, the same water keeps on circulating. Sometime in your life, you may have drunk water already chugged by a dinosaur. The connection between the oceans and the atmosphere can't be broken. What happens in one will always affect the other.

Water Cycle in Your Kitchen

You can make it rain, and you won't need your umbrella. First, make sure you have an adult around. Then, put a large hand mirror in the freezer until it gets really cold. Meanwhile, fill a teakettle with water, place it on the stove, and bring the water to a boil. Using an oven mitt, *carefully* hold the mirror over the kettle spout so that the steam hits it. What do you see? Draw the parts of the water cycle you observe.

Special Delivery

The oceans' salty waters, always on the march, carry precious cargo that affects our lives every day. By delivering warm water from the equator to the poles, ocean waters transport heat, making northern climates more bearable. By carrying nutrients from the seafloor to the surface, they nourish the plant life that is the basis of ocean food chains. And by pulling carbon dioxide from the air down to the bottom of the ocean, they slow down the dangerous buildup of greenhouse gases in our air.

Current Events

You can watch waves crashing on the beach, but it's offshore and under water that the real hustle happens. Enormous rivers of salt water—mightier than the Amazon, Nile, and Mississippi rivers combined—flow through the oceans on a worldwide trip that takes 1,000 years to complete. Other currents make shorter trips—traveling from sunlit surface waters down to the dark depths of the ocean floor, then back up again.

Thank Goodness for Central Heating

Massive currents flowing through the oceans act like baseboard heating for our entire planet. The Gulf Stream is one link in a larger Atlantic circulation system pulling warm water from the tropics into northern waters. Off northern Europe, the heat is released into the air, and the water cools, becoming heavier. Sinking down thousands of meters to the seafloor, the current continues on its journey back to the equator, where the cycle starts all over again.

Conveyor Belts? *Oui!*

The Gulf Stream and other currents are like massive conveyor belts moving a *lot* of warmth around the globe. The Gulf Stream alone delivers more heat to the North Atlantic than the output of a million power plants. This warming affects the entire Northern Hemisphere, but it's strongest in Europe, which is lucky enough to be near the current. Look at a map and you'll see that Paris, which enjoys relatively mild temperatures year-round, is actually very far north and should be as brutally cold as Labrador. *Ooh la la!*

Okay, Move It Along Now

What keeps the ocean conveyor belts moving? Heat. Heat fuels the winds that push surface currents around. And heat changes the density of water by making it lighter. Cold water is heavy and sinks; warm water is lighter and rises. Underwater currents move along because of the endless push and pull of these changing densities.

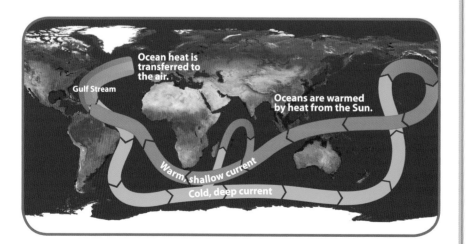

Ocean heat is transferred to the air.

Gulf Stream

Oceans are warmed by heat from the Sun.

Warm, shallow current

Cold, deep current

Imagine a water molecule riding this ocean current as it makes its way around the world's oceans—from the cold North Atlantic to the ocean floor to the warm equator and back to where its journey began.

You Can't Fool Mother Nature!

Salt water is heavier than fresh water. Warming temperatures affect this relationship. When seawater evaporates, water molecules float into the air, and the salt is left behind. The result? Saltier water. When more ice melts, extra fresh water is added to the ocean. So as global temperatures rise, tropical waters are getting saltier, while polar waters are getting fresher. In this way, global warming subtly alters the speed of the oceans' mighty currents. Scientists worry that eventually this will have major consequences for how heat is distributed around the planet.

Experts Tell Us — Naomi Leonard

Mechanical Engineer
Princeton University

When Naomi Leonard created her fleet of undersea robots, she took inspiration from the ocean. She decided they should act like a school of fish does—independently but cooperatively. When fish swim together, each one responds to what its schoolmates are doing. "If a neighboring fish is heading off quickly in a different direction, maybe an individual will follow, knowing that they have probably found some food or avoided a predator," she explains. "That's called responsive behavior."

Naomi designed computer programs so that her robotic gliders would have the same kind of decision-making abilities. They are fitted with sensors that measure all kinds of ocean conditions, such as temperature, salinity, and current speed. They travel as a coordinated team for weeks, covering hundreds of kilometers. Naomi stays high and dry in her lab, communicating with her robots via the Internet. "I grew up on the ocean, so it has been really exciting being able to apply what I do to problems in the oceans," she says.

CARBON-NATION

Heat isn't the only thing oceans soak up. They work as a planetary sponge for carbon dioxide, too. About half of all the extra carbon dioxide released into the air over the last 150 years has gotten stashed in the oceans. This is just another way oceans have buffered the effects of our bad habits. Without their ceaseless mopping up, carbon dioxide levels in the air would be nearly 500 parts per million—way higher than they are today!

Being There

What a Trip!

Carbon dioxide molecules from the air dissolve in cold seawater. Hitching a ride on those ocean conveyor belts, water rich in carbon dioxide heads down to the depths. It's on a trip lasting 1,000 years. But when the voyage reaches the warm surface waters near the equator, the carbon dioxide bubbles out of the water back into the air—like a carbonated soda going flat on a hot day. Scientists warn that as temperatures rise, it will be tougher for oceans to sponge up our mess. The warmer the water, the harder it is to dissolve carbon dioxide.

The Green Machine

Like their relatives on land, phytoplankton use photosynthesis to survive—using the energy in sunlight to turn carbon dioxide and water into food. So every day, billions of these tiny ocean organisms (right) are pulling tons of carbon dioxide from the water for a hearty lunch.

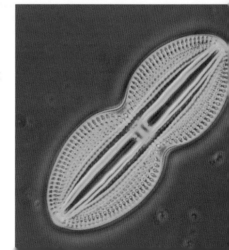

Phyto Link

Phytoplankton may be microscopic, but they're mighty. These tiny green organisms are the basis of most ocean food chains. They are invisible to us—except when they hang out in huge numbers. When phytoplankton multiply, billions upon billions of them make up vast pasturelands—or, rather, pasturewaters. Animals, from microscopic zooplankton to right whales, come to graze—the first link in the eating merry-go-round at sea. Phytoplankton multiply faster in a cool ocean. Earth-observing satellites in space show that the warming of the world's oceans is reducing the number of phytoplankton—hurting ocean food chains and speeding up global warming.

Billions of phytoplankton make the water milky green.

This tiny zooplankton glows in dark water.

A right whale flips its tail.

Experts Tell Us — Mike Behrenfeld

Biological Oceanographer
Oregon State University

Mike Behrenfeld loves going to sea. "Being out in the middle of the ocean is one of the most beautiful places on the planet," he says. Yet nowadays, Mike's research on phytoplankton relies more on satellites than on boats. "You can't see phytoplankton with your eyes, but you can see them from space," he explains.

Chlorophyll, the molecule inside plankton cells that converts the energy in sunlight to food energy, is green. So Mike can use photos taken by orbiting satellites to measure how much phytoplankton is in the water. Areas of deep green signal abundant phytoplankton, while bright blue means empty waters. Studying a decade of satellite images, Mike learned that rising ocean temperatures are reducing the number of these tiny organisms that feed zooplankton, krill, fish, and whales.

Three Cheers for the Phytos

Take a deep breath and say "Thank you" to phytoplankton. After all, they give us the air we all breathe. Since early life began on our planet, phytoplankton have been supplying Earth's atmosphere with oxygen and removing carbon dioxide from it.

Let It Snow

Most of the carbon dioxide taken up by phytoplankton is returned to the atmosphere when the plants die. But a lot stays trapped in the ocean when marine animals eat phytoplankton for dinner. When these animals defecate, or when they die and decompose, carbon-packed particles called marine snow slowly sink to the seafloor. There they get buried in sediment and stay put for thousands of years. If the carbon buried in sediment sits around long enough—say, millions of years—it can be compressed into oil and coal.

No Such Thing as a Free Lunch

When carbon dioxide gas in the air dissolves in the ocean, it combines with water molecules to form carbonic acid. But the oceans pay a price for this work. Soaking up all this extra carbon dioxide is making the oceans more acidic—already 30 percent more acidic—than before people started burning fossil fuels.

What the Best-Dressed Invertebrates Wear

The oceans also use carbon dioxide in chemical reactions that make one of the marine world's most popular building materials—calcium carbonate. Tiny corals use it to construct coral reefs. The most abundant type of phytoplankton in the sea, coccolithophores (below left), are protected by hubcap-shaped plates made of it. And countless ocean creatures, such as zooplankton, sea urchins, sea stars (below right), clams, oysters, and mussels, use calcium carbonate to build their exoskeletons or shells.

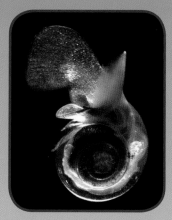

Butterflies in Danger

As the oceans become more acidic, it will get harder and harder for animals to build their protective shells and skeletons out of calcium carbonate. Scientists are especially worried about pteropods called sea butterflies. These beautiful, snail-like creatures are a key link in polar food chains. But their shells are very sensitive to acidic waters, so researchers fear sea butterflies could vanish within a century. And if the sea butterflies disappear, the many animals that depend on them for food, including zooplankton, salmon, and whales, would also suffer.

4 U 2 Do

Shell-Shocked

Even though the oceans are only a little more acidic, the change can make a difference. Drop part of a seashell into a clear glass. A piece of chalk will work if you don't have a seashell. Pour white vinegar (which contains acetic acid) over the shell until it's completely covered. Leave the shell in the vinegar for a few days. What do you predict will happen? Look at the shell every morning and evening, and write down what you see. After a few days, lift the shell out of the vinegar with a spoon. Place it on a paper towel, and press down on the shell with the spoon. What happened? Why?

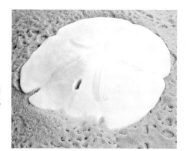

Being There

Sequestration Orchestration

Some experts believe they can use the ocean to sock away even more carbon dioxide. Projects are under way to store, or sequester, this greenhouse gas by injecting it deep into sediments on the ocean floor. Norway's largest petroleum company has buried millions of metric tons of carbon dioxide hundreds of meters underground. The Norwegian government is also exploring other ways to capture and store carbon.

[8] WEATHER FORECAST

Want to know how extra warming in the ocean far away could possibly affect *your* local weather? Take a long, hot shower. Standing there in the steamy fog, you can see for yourself how adding heat to water makes a big difference in the air.

Wetter World

The warmer the water, the faster it evaporates. And the warmer the air, the more water vapor it can hold. Eventually, all that water *has* to come back down to Earth—as rain or snow. Over the past century, the world has seen an increase in precipitation. The global forecast is for more intense rain, more often. But not everywhere.

Hot and Dry

Maybe where you live it has been getting drier. Well, guess what? That may be due to ocean warming, too! As precipitation patterns change, some areas, like the drought-plagued southwestern U.S., southern Europe, and sub-Saharan Africa, are actually getting *less* rain. So some parts of the world are in for more heat waves and prolonged dry spells.

Water in Lake Mead, on the border between Nevada and Arizona, has dropped 30.5 meters (100 feet) since the 1960s. See the white stripe? That's where the lake used to be.

Winds That Pack a Wallop

Many factors contribute to hurricanes and typhoons, but one important element is heat. Warm water provides the energy that drives hurricanes. Increasing the heat is like adding fuel to a fire. The result? Storms are becoming more powerful. As sea surface temperatures have risen around the world, so has the number of Category 4 and 5 storms—superdestructive hurricanes with sustained winds of greater than 209 kilometers (130 miles) per hour.

Experts Tell Us — Judith Curry

Atmospheric Scientist
Georgia Institute of Technology

When Judith Curry uncovered evidence that global warming is making hurricanes more powerful, she shocked a lot of people—including herself. Judith studies the interactions between the sea and the atmosphere that influence weather. Judith and her colleagues compared sea surface temperatures for all hurricanes since 1975. They used satellite images to judge individual hurricane strength. What they discovered opened their eyes. As water temperatures rose around the world, the number of extremely dangerous storms more than doubled.

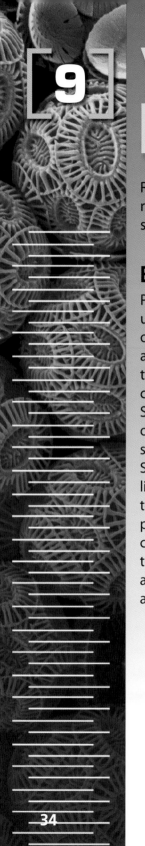

WORKING ON THE FOOD CHAIN

Rising temperatures threaten one of our ocean's most precious resources, phytoplankton—those tiny ocean organisms that support the vibrant web of life throughout the seas.

Eat and Be Eaten

Phytoplankton may be invisible to us, but they are the heart of most ocean food chains. In every ocean, animals of all kinds are drawn to the feast. Millions of tiny animals called zooplankton come to graze. Small fish come to chow down on the zooplankton. Then the small fish are eaten by larger fish. Seabirds and marine mammals like whales get in on the action, too. It all starts with minuscule phytoplankton, but the feeding chain goes right on up the line to top-drawer predators like tuna and sharks, polar bears—and us.

A laughing gull stands with its catch—a big butterfish (top). A gray seal basks on a rock (middle). Blue whales sing to each other to stay in touch (bottom).

Use the Elevator . . .

Like plants, phytoplankton need sunlight and nutrients to survive. The catch is, these vital ingredients reside in very different neighborhoods in the ocean. Sunlight is at the surface, but nutrients like nitrogen and phosphorus rest on the bottom of the ocean. There has to be a way to get these two together. There is—vertical currents, called upwellings. These upwellings act like elevators, delivering nutrients from the seafloor straight up to the sunlit surface waters.

Ocean organisms such as phytoplankton and these kelp live in sunlit surface waters.

Unless . . . It's Out of Order

When ocean temperatures get too warm, the elevators break down. Seawater becomes dangerously stratified, or layered—a thick blanket of lighter, warm water floats on top of colder, heavier water. This barrier blocks nutrients from getting to where they're needed. *Uh-oh*—system shutdown.

Mixing It Up

The mixing of nutrients by upwellings gets harder and harder as global temperatures rise and surface waters grow warmer. Tropical waters are particularly sensitive. Researchers say a temperature increase there of only 1 to 2°C (2 to 4°F) can mean a big drop in phytoplankton—as much as 30 percent. Why? The upward currents can't make it all the way to the top.

Rattling the Chains

When warmer waters prevent phytoplankton from multiplying, entire food chains get shaken up. Oceanographers studying the California Current have seen how warming temperatures can wreak havoc on a food chain. In the 1950s, these coastal waters supported a vibrant ecosystem, teeming with phytoplankton supporting seals, fish, seabirds, and whales. Since then, sea surface temperatures have risen about 1.5°C (2.7°F). The warmer water stratified more often and phytoplankton plummeted. This set off a domino effect.

Sea life out of sync:

1. Zooplankton in the California Current dropped by 70 percent.
2. Populations of seabirds, including the sooty shearwater, plunged.
3. Fish such as sardines seriously declined.

All Wrong for the Right Whale

Talk about rotten luck. They got their name because they were the "right whale" to hunt. So it's not surprising that right whales are now some of the rarest animals on Earth. Today there are only about 300 left in the entire North Atlantic. Experts have observed that when the mothers' food, zooplankton, is low the year before they give birth, fewer whale calves survive. If, as scientists predict, zooplankton declines as temperatures rise, this might be all it takes to send this struggling species over the edge to extinction.

This right whale mother swims with her healthy calf.

Attack of the Ocean Squishies

Of course, some animals love the heat. Jellyfish are thriving around the world—and causing jiggling heaps of trouble. Japanese fishermen have found their nets clogged with the gelatinous bodies of giant jellyfish weighing as much as 200 kilograms (441 pounds). And in the North Atlantic, vast jellyfish swarms are crippling already hurting populations of fish like cod by gobbling up their eggs and larvae. In 2005, Mediterranean beachgoers sweltered onshore as record numbers of stinging sea jellies made waters unsafe for swimming. And, guess what? They're ba-ack.

Sometimes schools of these giant jellyfish drift into the Sea of Japan.

10 SEA CHANGES

Our oceans are enormous. Because they are so wide and so deep, it takes a very long time for changes to spread completely through them. This means there's a delay in the ocean's response to the buildup of carbon dioxide and heat in the air.

In the Pipeline

The changes we see in the oceans now are really just the beginning. Scientists say that even if greenhouse gas emissions stopped completely today, the oceans would continue to grow warmer, creep higher, and turn more acidic throughout the century.

4 u 2 Do

Slow to Cool, Slow to Warm

Thermal inertia. Say what? This is just a fancy way of saying that water takes longer to heat up and cool down than air does. Take two similar glasses. Fill one glass with water and leave the other one empty. It isn't really "empty"—it's filled with air! Put both glasses in the refrigerator and wait 10 minutes. Then take them out. Which glass feels colder? Why? Do the same thing outside on a warm day. Which glass warms up faster? Why?

The Good News

Unlike a lot of problems in science, the really big questions about climate warming aren't much of a mystery anymore. It's not like finding a cure for a new disease or getting people to Mars. The fundamental causes of climate change are well understood. Better yet, most of the solutions are, too!

Our One and Only

One of the most important things scientists have come to understand is that our oceans can't bail us out any longer. Instead, the solution has to come from us. And the action experts say we must take is both very simple and very, very hard—we must reduce our greenhouse gas emissions.

From their salty mist and splashing waves to their curling currents and creatures large and little, the oceans are the lifeblood of our planet. So we all must take care of them to keep our planet healthy.

How can you help? By making smart choices about how you use energy in your home and school, and by urging the same from your family and friends, your town, and your nation.

atmosphere (n.) the mixture of gases (nitrogen, oxygen, and traces of others) surrounding Earth, held in place by the force of gravity (pp. 6, 8)

climate (n.) prevailing weather conditions for an ecosystem, including temperature, humidity, wind speed, cloud cover, and rainfall (pp. 6, 9)

food chain (n.) the path of food from one living organism to another in an ecosystem, showing who eats whom (pp. 25, 29, 34)

fossil fuel (n.) a nonrenewable energy resource such as coal, oil, or natural gas that is formed from the compression of plant and animal remains over hundreds of millions of years (pp. 9, 30)

greenhouse effect (n.) the warming that occurs when certain gases (greenhouse gases) are present in a planet's atmosphere. Visible light from the Sun penetrates the atmosphere of a planet and heats the ground. The warmed ground then radiates infrared radiation—heat—back toward space. If greenhouse gases are present, they absorb some of that infrared radiation, trapping it and making the planet warmer than it otherwise would be. (pp. 6, 7)

greenhouse gas (n.) a gas such as carbon dioxide, water vapor, or methane that absorbs infrared radiation, or heat. When these gases are present in a planet's atmosphere, they absorb some of the heat trying to escape the planet instead of letting it pass through the atmosphere. The resulting warming is called the greenhouse effect. (pp. 7, 25, 38)

habitat (n.) a place where individual organisms of a particular species live. It provides the types of food, shelter, temperature, and other conditions needed for survival. (pp. 13, 19, 23)

paleoclimatology (n.) the study of past climates (p. 17)

photosynthesis (n.) the process by which plants and other photosynthetic organisms use energy from sunlight to build sugar from carbon dioxide and water. As part of this process, oxygen is released. (pp. 9, 28)

phytoplankton (n.) microscopic organisms that live in water and perform photosynthesis (pp. 28, 29, 36)